TITLE

Health and Universe

XINZHONG DENG

BOOK and Volume

Title：Health and Universe

Amount of works: 14

Words：73,197

Page: 15.24cm*22.86cm 127

Article writer, cover photographer (12m pixel mobile camera), cover designer: Xinzhong Deng

Title of the cover: Health and Universe

FORWARD

Health and universe is still the nature phenomenon what human cannot control for themselves. Health and Universe is a book the author drop his view on the topic share with readers.

Private idea for only reference, where the knowledge should be taken from your textbook. If it's respond bad, it will be corrected or stop publishing as soon as it can.

Can not avoid mistakes and shortcoming want here your reader's excuse.

Content

~ 4 ~

1. Some Way to Change your Fitness

Each person's body, although it's nutritional status in fluctuations, basically are in balance. If you want to break the balance to change it better, exercise and diet are likely to achieve the purpose. Such as a body from sport looks stronger, because of the exercise can accelerate the tissue circulation, so that the body has more opportunity to bias the nutrition, lead the balance come to growth, to be strong. Some people try always to eat snacks, there may be used to eat less, get body in a state of starvation. When change for dinner to eat enough in nutrient rich, their behavior and physical situation may change.

Take more of the nutrition appropriately, can help the body to eliminate fatigue, as it can often be observed. For example, some of the old people go out by bus, come into joint pain, what from the deep contact with public items, consuming too much body energy, and some allergic constitution caused to inflammation because public hygiene. In addition to rest immediately, nutritional supplements should also be quickly selected to recovery the health. Inflammation above performance as itching or eczema, go to hospital may be appropriate, but if you choose some beneficial to resolve the body waste, food may be more simple and convenient. To someone health, can choose a liter of milk or pure fruit juice to try effect,

generally choose one, and can also use the two

together, different people may prefer different

one kind of the choice, if the body come to

heartburn or diarrhea shows it's in effect, and

there it may not be easy to bear, can

appropriately to reduce the dosage.

The method can also be used to lose

weight.

2. On Health New Way

Human body is a most efficient group to use energy, what can adjust its energy distribution for the needs of physical activity to reach in highest rate.

With the needs of energy of body it can automatically assign in the ongoing activities, most of the people do not doubt.

When running the digestion may partly rejected, the accelerated blood circulation is concentrated on the running, characterized by respiratory and heart beating faster, profuse sweating etc. Where the energy concentrated is where the part in moving, what concentrated in energy and nutrients because of moving get the chance developing. Sport as moving so can make the growth as result, shows as health.

In most cases, metabolism is the main way body consumption its energy. The consumption body maintain its exist is used to be to notice its heat dissipation and need to mobilize all the life to do, it's physical activated and would not be a express to most of the people. In test of penis erectile dysfunction, many people know the fact erectile in sleep and what happened almost every day, explain it's a physical reaction by the dream on sex and in fact it's a basic metabolism of its physical reaction result. Of the process metabolism large amount of body liquid flow through reproduction system to provide energy for its growth and cleanup the waste leading the way penis erectile.

Digestion, as the physical express of metabolism, need to concentrate in mind and energy and the sleep is the best way to do. If lack of sleep digestion capacity will be affected, digestive system diseases such as constipation may occur.

Adjusting metabolism by sleep body can mostly to do automatically. Some body or weaker easy to sleep at noon or afternoon is basically to satisfy the need concentrating on digestion after lunch. It's except of the morning because the digest ability come to most strong after whole night sleep and have enough power

to digest the simple breakfast.

Body energy under most circumstance of

constant body temperature to act its role quietly

and the body almost can not increase its energy

by heating, and if the body temperature rises,

its activities will also be effected. For instance in

hot shower amount of waste will be excreted

through skin by sweat and the waste in head,

when you blow strongly your nose, can also be

emission a lot in an unusual way. All above

come from the increased activity of   the body

tissue or body liquid when its temperature get to

high while the uneasy removed wastes get the

chance to come to the excreting organ and to be excluded by thermal motion.

The general state of the body energy would be understand by some phenomenon, that is a living organisms of its energy state would be observed. Body temperature is different in feeling from the general object, when touch in same the degree, feels more heat come from deep body, as life tissue of its energy movement can be resonate with each other and heat is issued by the tissue itself. When human semen has just shot out, it can be observed in the dark with infrared device, and will disappear a few minutes later, what tells the semen stay in a higher energy state in the body.

When the tissue or body liquid loss its activity,

means the life is no longer involved in the

overall activities, and the semen itself is

dynamic still, can be used in artificial

insemination, it would return to its energy level

as life free natural substance. Vitality within the

body should be a state of atomic thermal

motion what has the different digestive ability

from the substance outside the body. If the

state can not keep the vitality of life may be

affected so as to lead a variety of diseases. And

as the semen do not lost its activity to

reproduce it seems not a worry.

Health maintain by metabolism and

metabolism is a processes to redistribution the

energy between body and outside materials. Know the energy distribution and movement of the body can adjust the physical activity and therefore seek targeted health. Such as pure body tissue itself can maintain body temperature, when a large number of body wastes erosion, or energy is dispersed weakened, it will naturally diminish its physical function, try to pass the excretion, and even weight-loss, the body will be back to health.

3. Reduce Weight Ensure With Power

Reduce weight is always a practice related with diet and sport, what removing the heat from body, If understand obesity is not that heat accumulate to fat, loosing weight should go different way from usual,

Understand Obesity Correctly

There're reasons caused obesity and here the one come from the metabolism wastes what combined with fat together hardly to excrete. Obesity means too much fat stay in the body and the accumulation can not happen without reasons, when the wastes can not excrete, it store itself in the fat and shows weighted. The fat would not come in shapes when take high calorie diet, it's from the wastes of the food

body can not loose from metabolism.

Reduce Weight Need Plenty Of Food

So the precaution of obesity is lack or bad

of nutrition .Take enough food to supply enough

energy what do not reduce of the weight and

cause nothing of the fat, where the body grow

up to normal with the power.

When sufficient died do not effective to

develop to slim, it's from feeding habits.

Food in high energy and do not produce

wastes will give not obesity to body and keep it

health. When take less to lose weight, it will be

refused of the nutrition food above, and can not

stop the obesity. Take less food of its quantity

and the energy related quality, may shows slim

the results what come from the attention on

food from the quality and reducing adoption the

wastes from food as fat, give up the fat from

animals combined with wastes, give up the

wastes.

The effective way to reduce weight is the

demand of sufficient food supply, take enough

carbohydrates including fat and protein from

animals and plants.

How Movement Makes Body Health

Our health is a balance body against the

environment, some method to reduce weight is

effective and would turn back to fat in mindless

daily, health should come from the care of

environment and diet. Beside the daily life,

movement is not necessary because it waste

body energy. Body exercise strengthen the

parts fit and can grow up to balance with sports.

The essential of body exercise is to

stimulate with physical move, fasting the liquid

cycling and metabolism, make energy

concentrate on body or its parts, to make tissue

strong or developed. The movement do not

supply body the nutrition or energy and with

only sufficient food supply to effective. Without

nutrition the body exercise is not benefit.

Movement can reduce weight because it

strength the body circulation and fast the

metabolism what make the energy concentrate

to excrete the waste, what can hardly remove in normal.

The body energy should keep for essential metabolism. the symptoms related with age come from the weak energy, if it can get from only the way food beside of save.

Reduce Weight Need Body Power

One of the reason to obesity is lack of energy to excrete wastes or the poison settled down in body. The body power or energy is not only the fitness and strong, it's the wastes with high energy used by the body, and only when body get from nutrition as quality diet the energy can remove it to excrete.

Lack of nutrition as a health problem may come to be disease. Lack of energy need only food. Want to health at least not to be hungry and keep with a balanced diet, quality and quantity   of the essential nutrition supply make you out of the weight loosing problem.

Health Food To Keep Body power

Health is a state of body itself. Good nutrition will keep good body condition and express well. When diet and with only kind of food to excrete the wastes, body would not in its best in the normal. Excrete problem if not from tissue or incretion disease, it's from the hardness of the wastes, its accumulation cause body to obesity or disease and age.

The added nutrition supply solve the nutrition problem. Nutrition come into metabolism to make body health. With the nutrition body recover to its state and secrete hormone and digestive juice to excrete the wastes.

Nutrition can affect to body directly, where it react with wastes, change the combination bonds between wastes and tissue to make it excretion.

4. The Relation of the Energy between Food and Body Tissue Determine its Obesity and Health

Basically metabolism is a process body exchange it's energy with environment especially its food. Understanding the relation between body and food can make a better management on diet to reach slim and health. Obesity originally come from the accumulation of body metabolite, food in high energy and high quality can make body strong and health, and excrete wastes.

Digest of the food

Food digestion start from mouth, bites first then chewing, go into intestine and stomach and excrete off. Beside cutting and whipping, most of the digestion related with chemical reaction by digestive juice and enzymes , while absorbing and delivering by blood to everywhere of the body.

There's a different case may tell us the significance and result of digestion. When we smell of the food, excrete much of the saliva, certainly it's the first reaction of digest excreting in juice and in fact the results of the digestion come to the way: the molecular of the old tissue leave as endocrine or excreta while smell molecular in higher energy was decomposed and transfer energy to body.

Food transfer its energy to body

The health of body tissue related with the composed molecular energy .Body try always to keep its best energy level and comes to weak while live and life consume the energy. When lack of energy body waiting for the suitable energy molecular to feed himself, no matter what kind of state it is, gas, liquid and solid, all may be suppliers. So we can see the phenomenon: salivating by smell, because of energy supply the old body tissue come to be saliva to excrete. The reaction mostly express as urine and other kind of excreta. Someone diarrhea when drink milk because of lack quantity and quality protein. Some piss problem can not loose off take with only water, after have substance in high energy it shows work easy.

The excreta of body is often high energy wastes, what may be taken by body attached to its tissue inside or outside for energy support .Its accumulation may be the important precaution to obesity and disease and bad health. Because body uses the waste for need it can hardly remove out except body strong enough to have it off. So we can draws the conclusion, plenty of food especially high energy quality food is the premise and warranty for slim and health, as sufficient supply can make body develop to normal to strong, to have the power excrete.

What is high energy food

High energy food is not only protein and carbohydrate; it's the essential substance to keep body health. To disagree the way lose weight by less taking food of its energy there's the reason tissue cannot grow to normal or health when in hungry, no more power to remove wastes. If there's redundant energy left it would not save up to fat anyway, the fat in opposite of the reason as there's not enough energy to excrete the wastes accumulated in body.

The high energy food means here is the substance its molecular stay in high energy state, active to break the chemical bond leading the reaction as enzyme or vitamin etc. to specifically help body work in functions. In situation the same substance it can lead to exchange energy in atomic or molecular level.

With the example again for salivate by smell, mostly is acidic chemicals. Feeling of sour is basically a taste of hydrogen ion as hydrogen decomposed to only proton, known as smaller particle shows in higher energy to enter to physical exchange in digest.

As same as water, dilute water have more higher energy to loose thirsty because its molecular has a more closed structure, mineral water with high energy metal substance may not be absorbed by tissue because of its water energy level.

Take less meat should be take no more animal body wastes. Most animals meet the same problem can hardly remove wastes to deep and accumulate inside body, although shows not obesity, the environment or food conditions with hard wastes may saved. Meat obesity should be wastes of one kind of environment transfer to another one and can not excrete. If the food meat clear enough, as its body tissue of its life in higher energy, it's the high quality energy source.

More health is a choice body balance to environment to higher level

Body health can lasting in normal by ordinary daily food, good environment and personal hygiene. Because some of the body metabolism products would not the necessary part but was adopted by tissue to be energy support ,it can hardly removed from body, its accumulation will cause obesity included diseases. To keep body health clean and clear excreta is very important way.

As health knowledge spread and life level upgraded, essential nutrition satisfied everyone's need. The high energy quality food seems not some kind of foods, it's often different from its time and location where

produced .Sometimes someone fits to your

body and comes to normal after feed you

enough to its own level. The high quality food

made you body best should be a choice of

more better balance between your body and

environment and it's food.

5. Exelcymosis and so on

When pay attention on body there will be more or less can be found.

Inside body, especially the bone will express on teeth when its health coming bad. For instance feeling unsmooth of the teeth tells a demand the bone want excreta or nutrition to support its power, and can not satisfy the need to remove the waste. What almost the same the obesity happened to body. If body tissue can not excrete the waste off how fat a body we can understand.

The body energy use mostly on metabolism and its keeping, and the energy of environment can effect to the body directly. The first step of digest, the bite need energy support,

except the excrete of digestive liquid, the shape

of the cusp focus many of the body energy on

top to cut the food and make the food in same

state of the body of energy, so can easily go

through the digestive tract to digest without

foreign body sensation.

The consuming of the body energy what

can not supply in time should be the cause of

dental decay. As it's can not be escaped and

with lower food nutrition level, the common of

the decayed tooth shows serious of the

situation. For more on public place, the seat

and dinnerware is more easily take energy

effect with body than the food, leads crick and

arthralgia, what to the teeth the hard results is

dental decay. When in dental decay, even no

pain body get still problems as bad breath and

so on. (By the way bad breath related not only

with mouth, the unsmooth bone mentioned

before is also a source of bad breath, and

something less important as defecate may

come to be a burden to body liquid and express

in breath.)

Heterogeneous and stronger the infilling

make the tooth not decay more. Dental

examination in early time and fill the decay in

time can solve the problem. When it's coming

serious the tooth have to be removed off. After

pulling a tooth off you can find something secret of the body.

Though the wound on gum the healing will call whole body in reaction, consume lot of energy, need to rest a day home. Before you go to draw your tooth off finished an abundant dinner and prepared with fluid diet to supply your energy demand. When dental operation finished come home laying on bed, within about half of an hour you can feel yourself the reacting body repair. First the body circulation go faster up and start to heat up. As the body temperature is more higher than its before to the environment body feel itself chilling. It's especially hard on gum by the wound your blood is in running. While the energy is in consuming the body feel weak and can lasting hours or more. When the high temperature and faster fluid running stop together all your feeling recovery in a sudden body feel no weak at all, you can do anything as normal.

There's another phenomenon interesting. Generally we take food chewing without a sense of gum heat or cold. After a dental operation, chewing one side of your gum will feel obviously the heat, so we can confirmed that our teeth work alternately right and left side of the way, not many one noticed it.

6.   Understanding the Mysterious

Mysterious is a matter what can be feel only part and the sensitive matter express in different way in different wave band.

Mysterious things related always without vision or feeling, because of the existence of unknown, the phenomenon itself has been explained by many people, some of which is good enough close to the truth. If you also encounter the mysterious affairs, can you understand it?

As I see it is very simple, things is just a presence and the existence exist by a premise of feeling of the object, as long as we can feel, that exists. This is the simplest scientific justification on the exists, such as various ray photographs can explain the eye invisible matter presented.

And why some things can be felt but be seen? In addition to human photosensitive scope restrictions, the feeling and visual come in desynchronization is a basic characteristics a mysterious things exist, at the time things may be felt in some factors, what the target may be distance or perhaps close, and because of the various characteristics of the waves it can be felt, and while as the visual have to be seen in linear transmission, when it's out of the straight sight, or was blocked from transparent to the visible light, there will be only   feeling but a

view seeing.

Mysterious things act mostly the way in

such a phenomenon and there are examples to

the contrary too. It can only be seen but felt

nothing. Because the visual confirm the matter

more direct and external, it will not be thought

mysterious but an ordinary life, such as a

mirage, basically it's a invisible exist expressed

through the atmosphere refraction by the visible

light.

The body feeling beside visual and hearing

is generally a eternal feeling of body reaction

and seldom come in front of the face. When

come the striker would be frightened in strange

ideas. The support of body life essential is relatively stable and when it's excited in a sudden it will be uncomfortable or tense.

Here there's an example mysterious get with only feeling but visual, it's hardly discovered and would not be thought a mysterious.

Some flu like symptoms as throat inflammation related with certain habits. To see a doctor with some medicine would relieve the symptoms completely, and if it's observed carefully, we should be able to find the difference of the body before and after appear in public. If the body feel unwell, especially in throat swelling, giving arise, after eating in the restaurant, it should not use public tableware

when meals, so that the symptoms disappear

soon.

The things that happened in the premise of

the high restaurants, utensils were sanitized.

In normal circumstances it would not have been

thought possible to illness, and if it is the

reaction between different body could be go

through by some media as a object and the

existence of the utensils in some other forms

would not be touched in general ways, what

leads a

adverse physical reactions.

By the way there's a matter of life do not

belong to mysterious. Because of the

incomplete understanding of the human response to fluoride certain brands of fluoride toothpaste occupied almost all toothpaste market share. Some users allergy to fluoride does not indicate fluoride harmful, the fluoride toothpaste can be set on the shell. And if the fluoride is harmful the toothpaste on the fluoride additives should be end its use. Because there is no empirical research nor user complaints, some people may only sensitive to response to avoid of using of the fluoride toothpaste would be solve the problem. And for merely as allergic reactions, the brand toothpaste can not all be replaced by the senior fluoride toothpaste. The deputy response of the fluoride toothpaste is not easy to discovery, the user will first made

dry lips, and further more the back of the teeth

would be in web like ulcerate, some times the

pain is obvious。   When it was found, or there

are other symptoms of adverse reactions, it

should feedback to the manufacturers for

improvement.

Of the example you can not confirm the

reason of your symptoms and you can be sure

it's from a dental cream especially a fluoride

toothpaste and can take the action to remove

the sick. It's not a mysterious matter at all.

7. Physical Way to Digest

The article attempts to have a reasonable explanation on the digestion as an example from life and understand the way as digestion of physics. The physical digest may very little happens in the usual digestive system and when take place on the surface of the body it defense the body from outside invasion, what can make the invasion in complete annihilation.

What is the digestion?

We generally know digestion is food taken from outside into body of the life, to be decomposed by complex biochemical reaction by the digestive system. Such a digestion step, first by cooking and chewing physical methods of its mechanical way making the food cooked to degeneration, crushing, and then enter the digestive tract, through a variety of enzymes and chemicals, draw from the energy and nutrient out from the food molecules, trans them to the tissues, and finally excrete out the residue what cannot be further digested.

Digestion is focus on the absorb energy, and the use of food nutrients in body tissue to repair and growth. This paper wishes to explore the digestion step directly absorb nutrients or energy from food, not the mechanical steps of cooking, chewing, peristaltic etc. the normal digestive way.

Special example in physical digestion

Someone in rest, an ant like bug fly into his hole of the ear, with a sharp buzzing sound, the insect is instantly digested to disappear, and taste bitter in mouth. Then quickly went to the hospital for an examination of ear endoscope, the doctor denies abnormal with residue.

Discussion

This type of digestion, is very thorough digestion, organisms as food was completely annihilated, is a defensive bio-physiological reaction to body, make the body atmosphere automatically become an extension of biological individual. At end, the waste needs body excluded out, the most thorough digestive residual as inorganic chemicals and organic chemicals with the bond cannot be broken with the energy shows in bitter taste.

Digestion is generally carried out by smooth muscle in internal of the body, which has a variety of endocrine glands secrete enzymes and chemicals to break down and digest. The example of digestion occur in skin of the body surface, where the duct shape keep the skin get together with energy concentrating as internal body condition. Although there's not chemical substances from gland involved, the basic conditions is ready for the intermolecular energy shifting, in deeper the atomic level, take place the energy recombination similar as burning reaction. When reacting, the substance composed in molecules keep still of its shape, or come to different, to state of its energy changed. Means the atom structure keep still, the change of energy happened in atomic level, where the energy exchanges between food and body directly.

It's not necessary to know the reaction as nucleic or atomic reaction. The digest without added molecule join seems react in a great range of energy level. Moreover metal can be understood a big atom of its body, where the organic object has the same structure in comprehension. When digesting, its outer atmosphere can be easily absorbed, of the way as energy by the eater. By the fact the shape of

UFO, in some observation, only a surface as

extending of the body, easily breaks to eat.

If the above understanding physical

digestion was established, it is the important

step in vivo digestion should also be set up.

The only physiological chemical sign as taste

shows the level how the digestion can reach. In

theory should be the same as the burning left

only water, carbon and inorganic substances as

fundamental particles, and the bitter taste

seems to indicate that retained still the

presence of organic chemicals, the organic

substances go through duct skin absorbed to

oral cavity. Perhaps also the primary substance

particles from annihilation as intruders enter the body, to be captured into the bio chemical reactions, or, inorganic substances in forming have its own flavor.

The human digestive cavity should be easier come into the same above burning reaction, food molecular energy should firstly be absorbed, make the body strong enough to take part in chemical digestive reaction.

Bio chemical reaction, is the process the high energy bond of the food particles was broken and transfer the energy to the tissues, including energy transfer and the re combination of substance molecules from food to body. The digestion of the digestive system does not appear with physical digestion mentioned in this paper. Starting from broken food into the mouth, there's chemical reactions with endocrine, while in chemical reactions, the physical reaction is no longer come to run, in convenient the body use food material to repair or for tissue growth. When food in tooth cutting, the physical reaction comes, teeth pay the energy to the food to digest it, or absorb directly from food the energy to make food inactivation, get the teeth in different feeling. Teeth can play the effect to transfer energy directly.

The existence of chemical substances disperses energy of the body. Different chemical with different energy act in different physiological activity in interior of the body, the body tissue choose to make the chemical energy to fit itself. If the body can not break the chemical bond to use the energy, the food will eventually become waste to eliminated, and the high energy materials are usually not to be easily given up, can not excluded from body to the outside, so it will be changed to be energy dispersed excreta.

Conclusion

The phenomenon of life need a reasonable explanation, the insects annihilated in duct should be a physical digestion way to the body.

There's no chemical reaction in physical

digestion, but the material itself has changed,

by the way in very radical change.

8. Public Health Problems

When come to public places, the daily activities may have invisible health problems because of the cross use and the relevant situation need more study in explore.

If we know why sick, we will certainly take measures, and if we can not understand it in observation can only let it go itself.   If someone feel the pathogenic factors we did not know before, it should be a start in attention.

The sanitation quality of the public places is not always easy to understand, for example, the bus dizziness, restaurants to eat allergy, obviously looks clean, but why get body unwell?

Considered the issues, imagine the reasons,

with even the mysterious phenomenon of

physical explanation, can not show themselves

a solutions, finally take evasive measures to

solve the problems, as have dinner in

supermarket, fast-food restaurant or meal stall

where supply with disposable tableware,

walking to work nearby, the nearest shopping,

all can reduce a large part of the sub-health

problem.

In public places, especially the bus seat is

the main part raise adverse reactions to

somebody, in addition dizziness, swelling and

itching may come to contact part of the body.

The dizzy should be understood as the indirect effecting to each other of the human body. When people come to close, feel cool or heat as outside force distort or impulse on it. In effect to the object, the object will be changed to suit their own body force tendency or influence, to dispel the sense of distance, get the body comfortable. When people leave, the object itself keep still the changed forces influence, then in contact with any others, the human body and the object with the influence of their own have to deal in coordinate to consistent, what consuming with body energy, make it performance fatigue or dizziness. When the object powerful enough   wins the human

body, shows also dizziness.

To understand from another way, the

human body and the environment material keep

always in energy exchange. It's not easy to

keep an object maintain the average energy

level, it may related only to the contacted

person in front and when   the men physical

conditions in great big difference, perform in

discomfort because of the energy difference.

Swelling and itching is more likely related

to dirty, such as mildew,   toxic compounds and

microorganism etc. When see tableware and

seat clean clear with disinfection procedures,

one can scarcely imagine about with allergic

reactions. But a recent experience seems to deny that has nothing to do with the physical reactions.

Once had to go to a restaurant for dinner party, chose a just opened one to eat boiled dumplings. Scrubbing chopsticks with napkin before the dinner with the clerk looks affected, feel the only way to better hygiene. At first reluctant to touch the bowl, but the dumplings soup was attractive, in experience without exception fall to the allergy, for its start opening business, may come to different. Try in force to suck up the soup, a piece of hard thing run from the bowl edge into the mouth, down into stomach without a stop, certainly see nothing care about nothing , but the body reaction is not simple, feel suddenly a severely impact in rectum of the bottom. Feel no pain, but on the second day feel a lump or empty space plugging in the export to discharge, go washroom for two times to clean. Sometimes feel without the impact but keep still the lavatory reaction.

So should we know, the dining residue left in dishes as food or body material, would be the existence form can not be seen but are likely to be felt, they could pose a threat to human health. Without laboratory test, nor the reference material, but a French movie Wolf Empire where the dinner scene seems to be in conformity to the requirements of the

imagination. In the movie, what we thought the

putrefactive substances on tableware appears

in the face. The people in real life may not

necessarily not so, and the body itself may

adapt to their extend status. For body health,

make the right choice by the experience

knowledge, should pay attention to their own.

Certainly there's a lower possibility

explaining what we cannot escape as an

allergic reaction when get together. Different

man in different energy level, when contact

tightly will get in heavy interference, generated

to be allergen or microorganism, as outside of

the body different but similar the fertilization

between life, born in energy lower substance

with the problems as taste and infection and so

on. If it can be sure no one can live in avoiding.

9. Life And Universe

If life is a epitome of the universe then we can get from it a conclusion as a secret of the start of the universe. The article imagine of the start by life and understand of the life again by the universe start.

Of the hypothesis universe existed in only one object what kept in all the same outside and inside and in occasion there's a unbalance came to be leading to a result of the world present. The author imagine of the start and development by the reason force reacted with each other.

Science Need An Exact Precondition To Understand Matter

It's said One Drop Of Water To Know The Ocean means the relation of the matter, what the principle of the matter we know what's it being we get as the same of water and sea.

In an article of Scientific American Chinese edition 2006 Sept. issue on the topic of universe start tells there's a written blank of the developing of the universe from its start by the orange of human being in observation in various methods. Although science record need a life precondition of physical methods and an accurate understanding by thought will lead a proper view to search for at least the sideway of a substance and to be a basis of the imagine on

universe developing process.

The Original State Of The Universe

Universe come to be present there's

intelligent life in the earth beside sun, moon and

stars. What is the state of the universe before

the various come to be? It should be a

precondition of our imagination and is also a

precondition of the universe developing its life

in physics and in biology.

We thought the universe is in equality from

the very beginning, means there's only one

state without any move. Of a space without

moving the movement will be no counteraction,

the space is in the state of super conduction.

Super conduction is a state the movement was freeze to be zero or it's movement come to be no complex because of freeze. Beside the equality as a life exist there's no any other life, what life is a object with various ability and state of moving what its most complex by now is human body. What have been changed after the equality making the universe different from its original? We know there are many theories to describe include the big bang, what we imagine is familiar of the explosion.

The Equality may be or not exist at all, no matter what we thought it be to describe the beginning of the life in universe.

When there's a unbalance happened, imagine its an accident of the eternity still, it will lead a chain reaction of movement in an superconductive equality space.

The first force must be in direction and will be no stop when it start. A directing movement break the balance of universe and have all the universe concentrate to it, while it's superconductive, the movement is as severe as grand explosion.

The State Of The Movement Of Force

Movement has a property interfering, cumulating and being independent in its own moving direction, and the different kind of force can stay together. A very high ratio movement has the property as object to other relative weak force and there would be a reaction as an outer force act on an solid object.

No matter what it changed to be the first direction force, observed by any view, will suck any other exist around to its direction, what is the state we call as Black hole.

The second direction force vertical to the first direction force, what is the movement the surrounding concentrate to the space the first force go off. The movement of the force is against each other in horizontal and have an interaction with each other. If the interaction keep and do not escape at once its center will het strong of its density, what would be called as star. Of the background the vast of universe the movement will last long time and have a duration as the process of life of a star in death and live. It's also a possible cause the black hole exist in universe.

When the first direction force going on there's a effect generating to be the third direction force in wheeling or in opposite. It's exist in big or small ranges and in different direction cutting the first force in vertical and have the possibility to generate the independent star.

When the first direction force go through the second there has been ant force against to it, and what is nothing when in superconductive space. As ant force comes the movement in

first direction start to rebound coming to be one

a many one independent moving group like

moon or earth like planet. The kind of

movement can be referenced from the

moment-taking-picture as a liquid drop falling

into the amount liquid. A drop fall into a great

big liquid and rebound to be several liquid ball,

while the liquid ball go follow after the moving

liquid the ball would keep its state still. The first

movement fall into the first or the second

leading a blobbing reaction is the same of the

liquid will generate some force group for some

object.

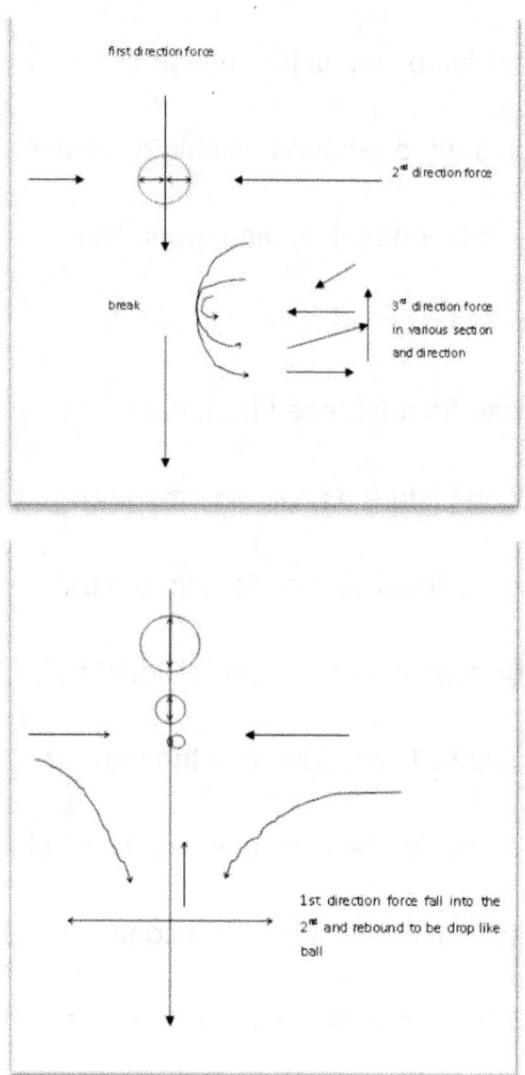

The reaction may be the essential way life

was created. What kind life come to be based

on the complex of the force around. The forces after its oscillatory come to be complex and stable, and full in universe. The force is not a problem as nothing but a time and condition to express. They all keep in its own characteristics and get together in the most save and stable way staying, and of course would go further more change by the changing of any of each force and finally to be other kind of life in universe. The intercourse in life is the most ordinary express of the process, what start from the most active and most small unit and developing to big and death because of the force of its movement, almost a repeat of the part of the universe from starting to ending.

While kind of force group developing the other group stay around to balance its stable exist and expressing in property of the substance as a expression of the diversify of the world.

The third direction force has many direction and section, come to be with even annular, what may be the predict gene universe generating in various. Three kind of force coordinate with each other will lead to be the shape of the galaxy we see as cake and helical.

After the starting of the first movement what will run without disappearing in zero or full speed, and keep in exist in different way to coordinated with others to balance to our viewed cosmic. The balance of the cosmic is still on developing. Any body is a result of the balance of the moving force, what the force is in changing to the time to break of the balance the body will be disappear to be other kind of more simple or more complex object. Because the vast of the universe the lasting of the balance can be   short or long, what is the process of life.

The movement will repeat of its oscillatory in its own ranges, each repeat have the same kind of forces generating and interfering, from small to large, keeping and filling of the universe of its range what is going on to developing.

The most complex of the movement

balance is human body. It's a reflection of the

universe balance.

Body is a result of the balance of the forces.

The force when observed among is in electricity,

and is in gravity when outside of the body.

Force can exist in most minimal body and the

most large range, expressed as electron and

the balance space. The smallest unit of the

movement can express in the same in different

balance space as proton, electron and so on,

and the large unit may be the space we live.

The magnetic field of the magneto electric

reaction may be one of its expression in focus

where the movement was picked up like water

is the electricity we use. When the movement or

space of the force was compressed it will

recover to be its original state in an open space,

what may be an essential level of our sensible

world in moving.

10. World of Substance is a Result of Movement

With many different theories the foundation of substance is the essential concern to human beings. This paper gives the views on the topic what helpful to understand the quantum mechanics and relativity theory. The universe movement force the space compressing to substance world, the compressing space has the substance property, the compressing space with substance property has the property of movement energy, time is the life what the movement keep in its still state.

The move of universe is the movement inside the universe

If movement belonging to a rigid overall, to universe, it's the movement of its body, seems as nothing to be. The movement in general is that happens inside the universe and get with the resistance from itself.

The pressure stress on forward from the movement is a premise to its resistance, when there's substance works it's the reaction. In original space situation, if it's not vacuum, there should be the substance related with the movement what is the premise for resistance of its original space material.

World of substance come from the resistance

These materials in hypothesis is originally assumptions and the action and reaction of movement make it change to different to be the current world.

When moving it compress the initial materials and when press to its extreme bearing end it start to give back a resistance reaction. The resistance should be a state of substance of its compressing space material and the movement stress express also the substance property in high speed move. Those two force react to each other in opposite direction, rebound and resist to its original direction force and rebound again, come to a circle movement of a spherical body.

The simple instance from life can help to understand the movement. A drop of water falling into liquid and return back to surface in resistance from liquid. If the drop fall in accelerate speed and the liquid start to move in same the direction with the stress from drop, the drop will keep still with its body in some where, what looks like the moon and stars in the sky.

The movement get with substance property because of the difference of density from moving

Move its location with the movement, for the matter in same rate or forced to the same rate to move the body, if it's not a rigid the different rate part will be compressed, when to

high different density and high speed the

moving body will change its state to be a

existing substance what separating part go

close together.

Even to extreme low density vacuum

space, in current view there's not nothing with

at least the various rays, will get with substance

property when compressed. And more to empty

space itself the outstanding movement will get

with substance property when in reaction.

The movements relative in moving, as in

premise speeding, when there's not a relative

movement inside the moving it keep in still.

When take a view to the other the power of

compression come from the rate diversity give

the object a substance property, means the

quickly change of the distance make the space

a compression substance.

The mass of a substance come from movement

The compressed substance come to be an

stable still exist, with its high speed movement it

has the momentum or energy of a substance.

When the moving body is extremely big, what

it's still inside, should be the state we stay as

substance world. When force on the substance

getting in a mini-relative movement or get in

focus of the first direction movement the energy

will release to its force.

The mass of the substance should be the energy related with compression from movement, the felt existing space compressed by movement should be the substance. The mass reflect the quantity of movement in a compressed space, or its compression rate of a space.

Movement is the change of a location, time state as invariant can help to understand the result of space compression.

The relation of a location in space need movement to reach and the movement need time, when the moving rate keep in definite value the unite move amount tells the distance and the space compression rate in high speed.

Of the inside of the compression space assume the distance do not changed, the whole moving body get in energy state the time start to be the life of its exist body. As the movement keep still in the state of a substance from its original moving state the movement keep still moving where the time with nothing changed.

When a movement changes its speed the substance as a combination of the related movement would come to break, while the physical body come to an end the movements keep to moving still inside the moving stream or original moving body.

11. Time Concept

The 4th dimension, one of the essential elements of nature, the time, what used to measure the lasting of a matter or a life of its age would not change of its unit in any way include even the situation light rate velocity. Mass and its emission is so different a matter we have the former in age as it's a life developing in different time duration, and we have the latter in spreading as it's an electromagnetic way to relive the mass of its record. The emission goes in ray velocity would not go through over the age, otherwise, mass and emission will be exchange from entity and electromagnetic to each other.

What's time?

Time, like the length and weight and so on existing in nature, is a essential element of our world.

Time as a measurement of lasting to a existing or length of life, rule with units to know the total volume, what the unit is unchangeable to its value. The total multiplied is called as age or the time of lasting.

The unit for weight is gram or Kg, for length is meter or km, and for time is second, hour, year and light-year etc. the developing of the unit let us to know the amount of a existing of its life or lasting.

Time measure with a rule the lasting of a matter or the length of life, the rule itself is constant in unit. The phenomenon time change with the movement in space is not exist, if there is, the unit of the time will different from its original value.

Space is the existence of self outside as the background, according to three dimensions of length, width and height to determine their location, Time is another dimension to determine the development of the object other than the location. Time will not be reversed, so the time of our lives can only be the time now present, to know the time in the past or the future, can do only in the present time, to understanding of the past recorded in the

media what processing as age.

On the lives of individuals, time is of age,

or process issues of its lasting, time like space

itself is only a existence of the background, and

can not be reversed.

The relationship between time and space

The time has not the properties above is

not the scale of a time, but a physical

phenomenon can be manipulated. For example,

in quantum theory, time as the space, of its

beginning and end, or the past and the future

can be simultaneously observed in a particular

location.

In general physics, a simple space can be

also observed and the time was in any case

can not go reversed back or observed both past

and future in same of the time with its material

state maintain in its entity. For this is the

premise of time and what can not go vice versa.

Because the results of observation of

cosmic physics is considered electromagnetic

wave movement, what go after a very long time

to be recorded as a imaging. The

electromagnetic waves move in the speed of

light, when approaching in or exceed the speed

of light, it will return to the past. This is only a

variety of outreach for the event, the event of its

physical entity is different and can not return to

the past, but there are past records of the

electromagnetic phenomena. If the

electromagnetic and material can be converted,

what keeps so many people want to know the

conclusion.

If the process can be converted into a

magnetic material entity, then the

electromagnetic process is the space, that is,

physical entities.

An indissoluble bond of time and space

may be a time come to a variable when a

movement go close to the speed of light, this

time do not mean time in the natural increase,

but to solely for the unit of time shortened or

extended due to the movement, the

measurement itself come to be variable lead

the time different.

The expand or contract of time maybe a misunderstanding of time and age.

The cosmic singularity, the so-called big bang of its position where the time and space unified, can there be time and space do not separate, and understanding the relation of electromagnetic phenomena to the physical world. Before the movement with distance appears, time and space stay together, when go emission in distance, the material world of its electromagnetic results, namely, the material results of the universe movement, start to change the mass entity to be different but retains the developing process in electromagnetic observations.

The so-called observe results of the material world, because of the time of life presented can show to observer in different stages. If the observer get the speed as the premise, can arrive at the material world of its different age. Where confused the concept of material world and the electromagnetic waves present observed. Is the observed electromagnetic imagine equivalent to the material world itself? At least not yet restored to modern means of the original physical form, the corresponding relationship between time and chronological age is definitely wrong.

Electromagnetic of its energy and physical

phenomena, indicating the loss of the physical world, or the changes of space because of the presence of physical force, what affect of the future of the universe, the latter shows the basic form of the universe, moderate and in personality, what present now can explain the past, the existence of life have a bright future, the former illustrate the seriousness of life because of the broader context of life related with movement. Also, because of the unity of commonality and individuality, to individual concerned, life is only our present landscape.

Whether the electromagnetic and material can be converted

Movement is a function of time and

distance, the movement in constant time units,

what the amount of distance changed by the

unit time, indicating the size of the velocity

where time is a speed constant variable,

distance come to difference while the velocity

keep still. The change of the distance can be a

prerequisite what the material world comes.

Before reaching the speed of light the faster the

relative speed of a static environment will be

observed as compressed state, and even

disappear; when static environment observe

moving body what may just invisible.

Relative to the speed of light motion, or

variety of moving exist forced to speed of light

together in sometimes, when moving in high

speed, it will be compressed together to constitute the material world itself.

To achieve the speed of light in motion, the objects start to keep relative still to the basic electromagnetic flow constitute the material entity. where various types of motion state can be observed in the location, can see the material world itself.

Movement has the speed of light, is a moving substance, movement carried out near the speed of light, there will be physical resistance, it will become the material in speed of the new energy level and need for greater acceleration forces to overcome the light energy level to be independent, its speeds different from the general .

Time can not be reversed

When achieved as long as the velocity of moving objects, it can appear in a relatively static position of the moving body, and reach the moving body's own material world. When achieved as long as the velocity of electromagnetic ray speed it can reach the material world of electromagnetic.

In the movement of energy level, the relative motion itself has the properties of materials. When a relatively speed more faster than the speed of the energy level, it will remain in movement to its advance state of the moving material, and maintain the basic energy level as ray speed etc.

Material express in time, come from the time go on the way to its different stages of life of the entity, time of the kinds of the states are the ages developing with time increasing. It's known as the age, or the processes of a certain stage, not just the time.

When a state of matter expressed in the form of electromagnetic waves, the age of the different positions moving in speed can be understood as a process beyond the time period, namely, reversed the time, what the time required for this process is still increasing, that is, the overall life-time of the material entity has increased, but the process went back to past. Therefore it is not back of the time, but the movement process.

If it's means time can not reversed, it is the observed entities, their time will not be back while the process of change is still increasing of the time, the movement of the flow constitutes the objects can be different in traced back or ahead because of the relative velocity, so you can be take to different stages of life because of the relative velocity.

So the direction of movement should be very important. The same direction of movement go into the larger stage of life, age in rapid movement make their early entry into the old, the reversed direction can allow yourself to enter into the life of the young stage. What kind of movement must be overcome the basic rate maintaining its own existence, the speed related with the speed of light in our world.

An accurate understanding of the young stage is, relative to the present, the age of some time past, in fact is an ages back to the past state at present time..

Summary

Time and age are two different concepts, the time to create age, age is but the living properties of material entities, what time itself has or had, so no further movement of the time, and was being recorded as amount of continuity of variety life due to the material changes in process. Because of the presence

of life processes in this context can go through

with a need of time, the process need a time, as

long as the process exists, it can be traced

back, or reversed.

The premise of a conversion between

material and electromagnetic waves is that the

speed gap between the two go decrease, when

the object reached the speed of light, it can be

reached in a relatively static state to the

material world originally owned by

electromagnetic movement.

12. Understands the Antimatter by the Quantum Theory

The property of electricity of a matter based on its wave nature of its direction and the direction of the movement denied the exits of a antimatter based on the electricity nature.

The wave properties of the particle

In quantum theory, particles and waves will be the same as an object to deal with, the so-called double phase of a particle. So we can see the basic unit of a substance would be also the motion, in the form of wave. The wave performance as a particle because it can be exist from the original move with the interference with others, what separate from the source and keep all its property to be observed independently. In another way to say part of the total movement changed its state and keep still the movement in the same way. Like those passenger in mobile devices, no matter what they do, they are only a part of the movement of the mobile, in same of the moving though they can do in complex in the mobile. The relationship of a man and a particle is that they both are passengers in mobile device, the particle moves the same as the man whom would not carry on the vast or background movement of the particle in his view, means man did not think him a particle in vast moving, what their exist nothing related with the wave

and is only a particle or substance.

How can a movement can be isolated from

source

The existing move, should be understood

as a movement of a wave with the nature of

particles. The kind of move is the essential

move of universe and was changed of its route

in reaction with other movement, what keep

always still to its original direction moving or it

used to move with a direction, lead the

movement in same power and same direction

concentrated to be a substance with some

particles.

The elementary particle of a material composition come from the basic movement what stress on the environment and the environment against to the basic movement, interference to each other to be a object in balance.

The electricity nature of the particles

Of the place we study, the state of a particle of its vast movement can not be seen and would be observed when in compression or in other media. For example the electricity is a compressed movement what fit to move in the media of its closer density or the particles in close property as conductor, and would be insulated to the substance with the particles formed in interference wave, because the exits of interference wave the movement of the electricity would be restricted what express as movement still as a tense in material surface as electrostatic charge.

Therefore, the electricity nature of the particles is a result the movement within a particle moving in direction.

The direction decides whether there is antimatter

The direction of a movement tells not only the movement go in a direction, it also means the focus of a movement, or its mass set in a particle. when in opposite direction the related property will be difference and can not create a particle.

To the structure of a particle, because the movement in direction and in volume, the opposite direction of a antimatter will not concentrate to be a particle, the antimatter in those opposite moving structure is not exist.

The opposite direction of the vast movement as interference may be a precondition to a antimatter particle. Because the interference wave is generated from the universal basic movement it can not fit in time and direction to basic wave the kind of antimatter can hardly be exist. If there is a movement same to the basic but in opposite direction, the inner move direction of the antimatter is opposite to the basic one and they cannot arrive in same of a mobile device in same time or stay in same a vast movement, so they can not come in same a place, if so, the interference of the two in opposite direction would form to be a kind of balance matter in the vast, keep in balance as star or atom and the interference based on the essential rule of the universe, when to balance they will move by the original movement of the universe.

The particle of antimatter in vast moving world move in opposite direction to right matter and the independence of a particle permits its moving in any direction, when a antimatter come in with right matter it should performs in

the same to right matter. When a matter come

from anti-direction vast moving to the opposite

one, because of its inertia it would be destroyed

as a explosion. For instance a man fly over

from a running train to another one would be

hardly reached, when the train go in extreme

speed will meet only fail.

13. On Speed

Movement in space with same speed shows as still, and as relative moving in space of different speed. Even ultra speed moving act as movement when there's no resistance, and when there's a resistance in the moving they'll come to interference to physical world or change the already substance to new combination.

Light rate is recognized as the fastest speed known to mankind and it's the premise of science, work and variety of human activities. If the velocity of light can go faster as we can move faster than the rate, life should be

different.

Imagine a reasonable state of the universe

The current universe, indicating that there

must be movement, no matter how movement

occurs, as long as uneven, there must be

relative motion, how much velocity should it be,

include all possible current movement speed,

and there should be a certain kind of movement

direction.

When go forward, encountered in resistance,

there may be rebound reactions, if there is

rebound, tells the motion space in substance,

such as liquid water droplets fall into water, will

pop up a lot of drop back, if the droplets at the

same time a certain speed with the water, the

rebounded water droplets will come to balance

with the moving water as molecule or solar

system to be.

By the initial velocity, the fast speed

movement move faster and get in higher

strength, the lower contrast, have the balance

system developing in different positions and

different sizes in space, then form the universe

we see as stars and sky. Because the speed is

sufficiently large, its development is broad

enough, what the difference is not necessarily

noticed related to these speeds.

$E = MC^2$ related issues

So the relativity theory formula is relatively

easy to understand.

$$E = MC^2 = m\,(at)^2$$

where mass as m is 1, t as time at the speed

of light is naturally derived in formula is 1, e as

the unit of energy is multiple of the force f and

its resistance in same amount in the time of its

unit as t = 1.

$$E = mc^2 = m(at)^2 = ma^2t^2 = m^2a^2t^2/m = f^2t^2/m$$

$$\text{as } E = f^2$$

The product of two forces described all the

power of energy, at last focus on to its internal

of the matter, originally the amount, finally the

same, should not change (in this case m=

$f^2t^2/E = f^2/E$ is the ratio between the two energies

is equal to one). When there is difference in

force, it will come to different material quality,

and should appear in different speed ranges,

and can continue to move forward. When use

force f to express the formula, time seems to be

variable, but the time should be decided by

force. If time can be changed, the shorter the

time, the energy will be greater, shows only

energy of its quantity when the material forming,

will not affect the movement itself.

When the velocity as light as c in a certain

rate with no prediction altering to be object, the

force f1(one of from f^2,as the first coming force,

cause the resistance as f2) will decide its size,

by the formula $e=mc^2$, c decide the e, where

without m before the touching, and when m

produced, m replace the e of its energy exist, or

e was storage in the m, m come in difference

because of c, where the energy come into

balance and without move more. In

accelerating movement where m is already for

the moving, the force f1 has not the prediction

change to be new matter as m, when f1 turn its

momentum to the given subject m, there's no

resistance f2 coming, may turn to go forward, or

start to change its state of the m itself.

The balances of fast moving come to

different position of the space. When a

movement with same velocity coming it will

come to be still within where move in the same

rate, or a relative movement outside the space moving in different rate. A relative moving inside come from a force on from its inside where the movement and strike force has the speed as light or ultra light in acceleration or deceleration, nothing related with the power form the substance or effected less. If there's object can endure light rate move strike and without go through and destroy it will gain the momentum to move.

Originally velocity of the moving nature in gradient

As speed is in variety, if you do not slow down and can stay still in a space or area, indicating that there is room for the existence of different speeds, and the first place keep you still will see you in motion, on relative of the speed there's no difference when it's fast and low. When fast, or fast enough to compress space into a stationary state of matter, as long as you can observe, the situation remains. If you can not do it, simply because the vast space caused by rapidly moving can not put the two motion together for compare.

At present concrete examples should include other stars around the Earth at an acceleration away from our observation, if the current observed star separating speed in difference, indicating that the material world of the universe itself is in uneven speed, fast motion can be seen accelerated run away,

relatively slow motion observed the same

because the Earth itself is rapidly move off.

In addition, as the satellite launched did not

slow down after liftoff, it can stay still in orbit,

indicating its new location in different speed

from where it run off, a state as the same rocket

accelerated, and so should be sure that the

space around the Earth is in different speed to

keep still. Step out of gravity, as to make an

object with a destination speed, simply is to

adjust its motion inertia to fit the new speed to

keep still.

Gravity should be characteristic of a

substance in moving. When the directional

move form the galaxies, its direction facing the

center of balance, so the ground or around

shows as gravity to the center.

The possibility a fast move changing the

material and the life

From the earth the other stars observed go

far away in acceleration, is the size of the

velocity is in obvious difference? It should be

different, that's tell the universe run in some

direction and with gradient. In this way we can

imagine the move of the universe very

differently.

As assumed in the first paragraph, the

material structure itself is outcome from the

directional movement, maintaining the inertia to

holding its position, determine whether its state can be kept, that is, whether can be itself. If the inertia adjustment has no effect on the material itself, the Voyager can keep its own life while flying out of the solar system, to finish its established mission. Intelligent life as human kind should be the same.

Force can change the location, and also to the inertia, and what to their impact on the structure of the material balance system? Substance of a balanced system can keep in exist independently, but will change shape when force is strong enough, in order to achieve a new balance. Imagine loosening elementary particles in a particle accelerator, how can it be regrouped, indicating the

changing possibility is very large. If the state of

particles eventually disappears, means the

motion itself determine the instructions and

generating of the substance.

When the acceleration is less than the rate

formatting the material itself, can speed up its

movement, and when greater than the rate of

formation, will like a bullet penetrate the car,

forward with itself, if the car can not be

penetrated, it will be compressed, this perhaps

the situation occur in super-fast movement.

Certainly if not for the fast-moving the

objects get in some distortion to the observer,

but because of fast moving objects has its own

absolute deformation, perhaps the rapid games can make objects deviate very far from its own inertia place, meaning its position of the movement has to change a lot of inertia to adapt to a new location, basically it's a strong external strike, should be devastating, not just deformation. If the aircraft in flight can change its inertia, or accelerate by itself to adjust its original force forming the substance, make it stay still in a space of different rate as the striking force running to be, it will not be deformed.

If the purpose of the movement change the inertia itself, the substance of its structure will be changed naturally after with, the movement will not get in deformation problem. If the object structure can withstand squeezing force generated by the movement, the moving object could not have its body deformation.

There's still the possibility the objects come to be shorter. We thought tight solid objects, including human body, in some people's eyes in different inertia velocity is just a bunch of loose material collection, there were big gaps allow some moving body or organism go and out randomly, when running fast, the gap will be shortened, so as to the object. And whether the state of life will be changed, there may be different possible. It may be a key to the move limit. When the resistance of the gap is destroyed and leading the material of its

constituting or the object come to different, the

moving object naturally come to the moving

compression space to generating state to new

material, the initial raw material can certainly

generate to different, and come to be new in

balance of the new velocity.

14. Special On Quantum Of The Book

You Reading

The book Health and Universe has already come to an end and when the author take part in discussion by WeChat keyanquan, deal it again. Here below review with digest for reading.

Shell and Squirrel is featured products of Amazon and Nature and Science can bought here too. Health and Universe get from life and explains in different way with hypothesis how the substance world comes from, what could be a prediction to understand quantum physics. If the present conclusion by observation is not wrong, it should come to be the same in study

even there was not a one open to tell now.

The book just classify to life science,

universe on its topic take the subway to explain

the life of its helpless. Profound understanding

the life but not deep into the quantum physics in

reviewing detail, every theme stand in

independent without relationship, although

some discussion may on the example as

quantum. Man knows in advance before the

devices.

The quantum discussion start from the

difference between   calculation and computer

operation. Calculation pick the data from life or

a process then go through some rules to get in

conclusion what was take to the process to

work. Calculation process is different from the

life process, without the prediction of gravitation,

is not a quantum state as computer operation.

The wave from quantum based on substance

and get the property of gravity. Handle quantum

wave is not as simple as change life with

calculation.

Computer operation is running with

machine as substance, a state of collapse of

quantum. The keyboard of the question is

gravity. The movement, when make up to be

substance, press together to be gravity. Gravity

is only the movement changed its running

direction to core of a substance. If a movement

go through the space with concentrated

movement, may it will shorten the time off, what

an example wonder the wave state exist in

quantum substance.

Gravity is carefully concluded at last of the

book by author, and it's helpful to understand

the quantum physics. Wave and substance has

a distinct interface and its movement

connecting to each other still, what leads the

discussion in quantum about wave particle

duality. If the conclusion is right may the

dimension in physics can be simply to present

four now.

From duality, substance is the result to a

coherence as collapse of a quantum. The

difference of the substance world come from

the difference of its wave character. The

evidence of the duality coherence seems to

search the gravity difference between the target

particle and its condition. Imagine the

conclusion but hardly to embody. Quantum

physics should finally find out the wave content

of a substance of its wave coherence. By now

only human body can feel the exits out of its

body surface, what may be the state of gravity.